SEA CREATURES & ANIMALS

IN HONG KONG *AND* FROM AROUND THE WORLD

海洋及陸地的動物朋友

SEA CREATURES & ANIMALS IN HONG KONG AND FROM AROUND THE WORLD

is published by Parkland (Hong Kong) Limited, a consortium of
Sino Land Company Limited and Empire Group Holdings Limited.

Designed & produced : Parkland (Hong Kong) Limited

Descriptions & illustrations : Professor Brian Morton

ISBN 978 - 988 - 75921 - 0 - 5

《海洋及陸地的動物朋友》是由信和置業有限公司和帝國集團控股有限公司
間接持有的合營企業栢聯 (香港) 有限公司出版。

設計及製作 : 栢聯 (香港) 有限公司

文字介紹及插圖 : 莫雅頓教授

國際標準書號 : 978 - 988 - 75921 - 0 - 5

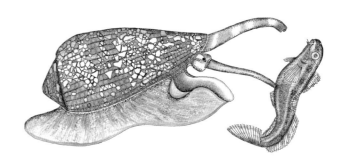

SEA CREATURES & ANIMALS

IN HONG KONG *AND* FROM AROUND THE WORLD

海洋及陸地的動物朋友

A CHILDREN'S GUIDE BY PROFESSOR BRIAN MORTON

莫雅頓教授的兒童圖書

CONTENTS
目録

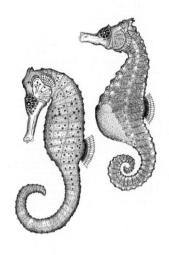

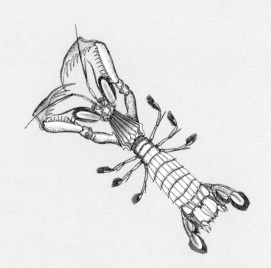

FOREWORD BY DR. LEUNG SIU-FAI, DIRECTOR OF AGRICULTURE, FISHERIES AND CONSERVATION

前言 — 漁農自然護理署署長梁肇輝博士

———————

It was in 1986, when I read for a postgraduate degree in The University of Hong Kong, under the supervision of Professor Brian Morton, that my lifelong friendship with Brian began. From an ordinary mentor-mentee relationship to a strong bond of friendship over the years, I have much to thank him for. It is therefore my great pleasure, both as his past student and a close friend, to write a foreword for his latest piece of work.

Brian has been termed, very appropriately, "Hong Kong's father of marine conservation" in view of his enormous contributions to teaching and research in marine sciences, in nurturing generations of marine biologists, and in pioneering and fostering marine conservation and public education over his 33 years of dedicated service, continued uninterrupted despite his retirement in 2003, in Hong Kong. Well known for being a prolific researcher and a great scientist,

我與莫雅頓教授維繫一生的友情，始於1986年，當時我在香港大學跟隨他修讀研究生學位。歲月悠悠，我們由普通的師徒關係進而成為知交，他的深厚情誼，我一直銘感於心。他既是我的授業恩師，也是我的良朋知己，可以為他的最新著作撰寫序，我感到萬分榮幸。

莫雅頓教授被譽為「香港海洋保育之父」，這稱號實至名歸。他在香港孜孜不倦地工作了33年，於2003年榮譽退休後，工作熱誠卻依然未減。莫雅頓教授在海洋科學的教學熱忱深受愛戴、學術成就超卓，多年來培育了許多海洋生物學家，他也一手開拓和鞏固了香港的海洋保育和公眾教育工作。莫雅頓教授無疑是一位傑出學者、

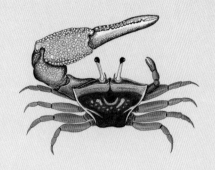

Brian was also a highly gifted biological artist, as seen in his scientific drawings and illustrations. He had learnt the skills to draw accurately when he was still a teenage student, volunteering to help his zoology teacher with pencil drawings as seen through a microscope. Brian's years of relentless and diligent work further honed and perfected his techniques. Coloured drawing, for example, was self-learnt after his retirement, built upon his excellent previous experience in monochrome drawing.

The forty drawings in this book, mainly of marine creatures that occur in Hong Kong, but also other animals of special interest to Brian or of ecological importance in other parts of the world, are splendid demonstrations of biological details and artistic beauty. Preparing these drawings and the associated text descriptions demanded a rare combination of talents in both science and art, and Brian was one

蜚聲國際，不過他也是一位才華洋溢的生物學畫家，這從他所發表文獻中的畫作及插圖可見一斑。莫雅頓教授早在少年時就學習繪畫，源於他自薦協助動物學老師，用鉛筆準確臨摹顯微鏡下的影像。經過多年的辛勤鍛鍊，他的繪畫技巧得以與日俱進，漸臻佳境。彩色繪畫也是他退休後，憑藉他過去豐富的單色繪畫經驗，自學而成。

這書收納的40幅作品，大部分是本地的海洋生物，也有些是莫雅頓教授感興趣或其他地方的生態重要物種。這些畫作糅合藝術美感，細緻入微地呈現出不同生物各自的形態特徵，實屬美妙佳作。繪畫及為每幅作品撰寫精簡描述，必須同時俱備深厚的科學及藝術造詣，莫雅頓教授正是極少數兼具這兩種

of the very few who possessed both. I sincerely hope that this book, highlighting the rich biodiversity of Hong Kong and worldwide, will help arouse the interest of the young generation to learn more about our vibrant marine life and diverse habitats, to care for and cherish the ecosystem services we now gladly enjoy, and to take active roles in safeguarding the sustainability of our marine environment.

To our great sadness, Brian passed away on 28 March 2021 at the age of 78. While we deeply grieve the loss of Brian, we should also celebrate his life which is full of wonderful memories and accomplishments. The publication of this book is timely, for it shows how Brian's work can continue to influence others, not just through the academic excellence that he had attained, but also through his vivid, colourful and captivating drawings. As his first ever book written specially for children, this latest addition would help attest to the fact that apart from adults

迥異稟賦的優秀人材。我衷心希望這本強調香港及全球豐富生物多樣性的著作,能引導年青一代去認識香港所蘊藏的美麗海洋瑰寶,關心和珍惜各樣我們樂在其中的自然遺產,並且積極參與、維護及推廣海洋環境的可持續發展。

莫雅頓教授在2021年3月28日不幸與世長辭終年78歲。痛失良師益友,內心難免傷感,然而莫雅頓教授已走完了精彩人生,留下了段段美好回憶,我們也應該為此而欣慰。出版這書正好讓我們可以向這位傑出學者致以深切的敬意。這書清楚顯示莫雅頓教授不僅以他卓越的學術建樹惠及同儕,也可以創作出生動、斑斕、賞心悅目的畫作供大眾觀賞。這書也是莫雅頓教授第一本特別為兒童而預備的著作,是一個很好的明證,

like students, scientists and conservationists, children and the young, particularly those eagerly seeking and pursuing their dreams, could also learn and gain from Brian's legacy; and perhaps more importantly, be inspired by the legendary life of "Hong Kong's father of marine conservation".

Dr. LEUNG Siu-fai, JP
Director of Agriculture, Fisheries and Conservation
HKSAR Government

說明並非只有大專學生、科研或保育人士等成年人才有機會受惠於莫雅頓教授的非凡成就；兒童及年青一輩，尤其是那群正熱切追求理想的尋夢者，都同樣可以受益。不但如此，這位「香港海洋保育之父」傳奇一生的豐富歷練，實在足以啟發和激勵一眾年輕人的青春生命。

梁肇輝博士, JP
香港特別行政區政府漁農自然護理署署長

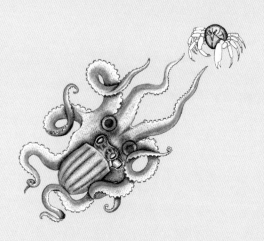

MESSAGE FROM MR. LAU MING-WAI, CHAIRMAN OF OCEAN PARK HONG KONG

香港海洋公園主席劉鳴煒先生的話

In Hong Kong, conservation and sustainability have come a long way. Hong Kong may historically have begun life as a trading outpost that had to be developed at all costs. But today it is at the forefront of the debate and research on how best to protect the environment and sustain biodiversity. The public is ever more aware of the impact their lifestyles have on the natural world and individuals are constantly making changes with conservation in mind. These achievements did not happen overnight. They are the fruit of decades of research, advocacy, and public education.

While no institution in Hong Kong can be credited for single-handedly bringing about these achievements, Professor Brian Morton was instrumental in all of them. Professor Morton was one of the pioneering marine biologists in South China. At The University of Hong Kong, he trained 39 PhD and 23 MPhil students. Many of his students are now in leadership positions in government, academia, and NGOs. He founded the University's Swire Institute of Marine Sciences and Hoi Ha Wan

於過去數十年間，香港在促進環境保育及可持續發展方面已有所進展。時至今日，我們已積極研討環保議題及維持生物多樣性，社會環保意識普及，大眾亦更了解到現在生活方式對自然環境的影響，並開始在保護環境的前提下於生活習慣上作出改變。這些成績並非一朝一夕得來的，而是經過數以十年計的研究、宣傳推廣和公共教育才取得的豐碩成果。

雖說這成果並非由一個單位獨力促成，然而莫雅頓教授在促進各項目成功中扮演重要角色。他是南中國首批海洋生物學家，於香港大學期間一共培訓了39位出類拔萃的博士畢業生及23位碩士研究學位畢業生，其中不少學生現已分別在政府、學術界，以及非政府機構擔任領導要職。莫教授亦是太古海洋科學研究所和海下灣海洋生物中心的創辦人，曾積極參與設立米埔自然保護區、海岸公園及海岸保護區。

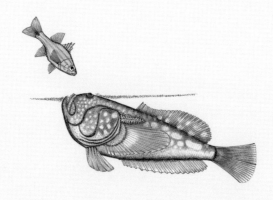

Marine Life Centre. Professor Morton was also deeply involved in the creation of Mai Po Nature Reserve and Marine Parks and Marine Reserve.

As a dedicated educator, Professor Morton understood the importance of getting the public involved. He once said "you cannot effect either conservation or environmental protection unless you have an educated public". His lifetime of work attests to this belief. It is all the more touching that this book is one of the final projects of Professor Morton. We at Ocean Park and in Hong Kong are forever grateful to Professor Morton — Hong Kong's important marine habitats and marine biodiversity can be protected and conserved for our future generations because of him.

Mr. LAU Ming-wai, GBS, JP
Chairman of Ocean Park Hong Kong

莫雅頓教授是一位充滿熱誠的教育家，他充分了解公眾參與對環保的重要性。他曾說：「除非社會大眾都掌握環保基本知識和概念，否則你將不能有效推動任何保育或環保措施。」他畢生的工作一直秉持着這個信念。這著作是他晚年寫成的其中一本作品，深刻體現他的科研精神，尤其讓人動容。全因莫雅頓教授努力不懈的堅持，才能為香港下一代保育重要的海洋生態環境和生物多樣性。我謹代表香港海洋公園以至全港市民對此致以最衷心的謝意。

劉鳴煒先生, GBS, JP
香港海洋公園主席

INTRODUCTION

簡介

———————

It is now seven years since I published a book (in 2014) of some of my animal and plant drawings as Tales & Drawings of a Life by the Sea. This is out of print now but is freely available as an e-book with Apple iBooks, Kobo, Smashwords, Barnes & Noble and Amazon Kindle. It was produced because, since taking early retirement from my Chair of Marine Ecology at The University of Hong Kong in 2003 and returning to the United Kingdom and home town of Littlehampton on the West Sussex coast, I began to re-discover the relaxation afforded by drawing. And, using the lifetime of experience gained by researching and describing the marine life of Hong Kong and elsewhere, the conclusion reached was — why not continue to pursue this pleasurable relaxation in retirement? And, in the hope of improvement.

By coincidence, the above book came to the attention of Nikki Ng of Sino Group, a company then planning but now creating a new hotel associated with Ocean Park in Hong Kong. Nikki visited me in Littlehampton and described what she would

我於2014年將部分動植物手繪作品結集成書，名為 Tales & Drawings of a Life by the Sea，如今已有七年。那本書現已絕版，但可於Apple iBooks、Kobo、Smashwords、Barnes & Noble 及 Amazon Kindle 等各大電子書平台免費下載。我於2003年提早退休，結束在香港大學海洋生態及生物多樣性學系的教學生涯，返回我的家鄉 — 位於英國西修適士郡 (West Sussex) 的海濱小鎮 Littlehampton，自此再次重拾繪畫的悠閒樂趣，不知不覺完成了這批作品。於這段期間我決定將我畢生研究及敘述香港及世界各地海洋生物的經驗與大家分享，享受退休生活的同時繼續繪畫，也希望讓我的畫功更上一層樓。

機緣巧合之下，這書引起了信和集團黃敏華小姐的注意，那時信和正計劃發展一個與香港海洋公園有關的酒店項目，現已動工興建。她特地到 Littlehampton 找我，誠邀

like me to produce — forty new drawings of marine animals mainly focused in Hong Kong. They would also be used to illustrate a new book — this one — of my new drawings. How could I say 'No'?

In early 2019, therefore, I began to search through all my old photographs of plants and animals that I had come across in my many years of undertaking research on the marine life of not just Hong Kong but other locations that I had visited and, often, undertaken research at. These images became the basis for the drawings that I would slowly begin to recreate on paper with ink, pencil and brush.

Of the forty drawings of species reproduced herein over half occur in Hong Kong but, in reality, they are also elements that occur throughout the vast Indo-West Pacific region on Earth. From tropical Japan in the north to Northern Australia in the south and from the Indian Ocean in the west to throughout

我繪製40幅全新海洋生物作品,主要源自香港的品種,而這些畫作將用作這本新書的插圖。盛情難卻,於是我就一口答應了。2019年上旬,我開始從我的動植物舊相片中尋求合適的參考材料,當中包含我歷年來在香港及世界各地所接觸的海洋生物,尤其是經本人研究的品種。其後,我以這些相片作為藍圖,利用鋼筆、鉛筆和水彩畫筆,逐一將相片重現於畫紙上。

畫作裡40個品種中,近乎一半源自香港,而其實牠們亦遍布印西太平洋地區,包括位於北半球的日本熱帶地區、南半球的澳洲北部,以及橫跨西邊的印度洋以至物種多樣的東亞水域,乃至太平洋群島及環礁。其餘畫作的選材則不外乎兩個原因:具保育價值和個人興趣。前者突顯某些物種在保育的重要性,如兩個中國大熊貓品種;後者則表達

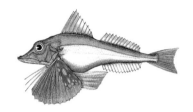

the diverse waters of East Asia and spreading outwards into the islands and atolls of the Pacific itself. The remaining drawings are of species chosen either because of their conservation importance, for example the two species of Chinese "pandas", or simply because of the interest they had for me at the time of drawing and further because they illustrate some feature of interest or, again, have conservation importance. For example, the ice bear, or polar bear, captured as a photographic image while on a small vessel moving along the eastern coastline of Greenland — its remotest and least populated region. And, the little egret, once well known to me at the Mai Po Nature Reserve in Hong Kong but also now (since 1996) resident throughout Great Britain. All of the drawings therefore are a personal memory, either reflecting my thirty-four year's residence in Hong Kong or adventures elsewhere either during that period from 1970 to 2003 or subsequent to my departure from Hong Kong.

我感興趣的範疇，或一些特別元素，當中也包括保育價值。舉例來說，我曾到訪最人跡罕至的格陵蘭東部海岸，在小船上沿著海岸線遊覽，沿途拍攝到北極熊（或稱冰熊）照片，而這本書介紹的另一種動物小白鷺 — 以前我只會在香港米埔自然保護區看到牠們，但牠們自1996起遍布整個英國。因此所有畫作都盛載著我的個人回憶，一是見證我在1970至2003年的34年間在香港居住及探索世界各地的所見所聞，二是我離港後的親身體驗。

香港漁農自然護理署署長梁肇輝博士是我一位認識最久的知心好友，熱心為我提供一些本地動物相片作臨摹之用，當中大多是魚的相片。我十分感謝梁博士所提供的

Some of the Hong Kong animals, notably the fishes, were drawn from photographic images kindly sent to me by one of my oldest and closest friends, Dr S.F. Leung, Director of the Agriculture, Fisheries and Conservation Department of the Hong Kong SAR Government. I have a great deal to thank Dr Leung for in other respects too, but that is a different story. Another close Hong Kong friend, Dr K.F. Leung of the Environmental Protection Department again of the Hong Kong SAR Government, has also been a source of inspiration for the drawings contained herein.

A little more explanation about the drawings is needed. In essence, they are simply stand-alone images of our, mostly, marine creatures that we share this planet with. They are also some of our more obvious creatures that we see when visiting either a beach or a market fishmonger's stand. They represent the larger elements of marine life and omitted are the millions more numerous

協助，而他對此書的其他方面也貢獻良多，在此未能盡錄。香港環境保護署的梁劍峰博士亦是我另一位好友，給我的畫作帶來源源不絕的靈感。

我亦想藉此解釋畫作背後的一些理念。這些作品基本上是一幅幅獨立的圖片，呈現出在這個地球上與我們共處的海洋生物。牠們代表海洋生態中的體積較大及較常見的品種，平常我們也會在海灘和街市魚檔找到牠們，然而海洋中仍有數以百萬計肉眼難以觀察的微生物，在此未能盡錄。

每幅畫作都附有動物俗名及學名。俗名會隨著物種所在的國家，甚至地區而有別，例如 *Plectropomus leopardus* 的英文 coral trout（東星斑），擁有近119種已知俗名，均由西太平洋

microscopic creatures that one never sees or, if so, rarely so.

Each drawing is accompanied by a common name and its corresponding scientific name — the latter provided because common names attributed to a species vary from country-to-country and even region-by-region. For example, *Plectropomus leopardus*, the coral trout in English, has one hundred and nineteen known names given to it by the people throughout its large Indo-West Pacific range who fish for it and eat it. As a consequence, the original Latin name given to it, in this case by the French naturalist Bernard Germain de Lacépède in 1802, is hugely important because if I were to consult, say, a Malaysian fish specialist and ask about the incidence and status of 'coral trout' in his nation's waters, he would not know what I was talking about. But if I asked the same question using the name *Plectropomus leopardus*, he would know instantly what I was referring to and would be able to accede to my request for information about it. And, thus,

地區的漁民和居民所命名。因此,由法國自然學家貝爾納·熱爾曼·德·拉塞佩德 (Bernard Germain de Lacépède) 1802 年給予東星斑的拉丁學名更顯重要。倘若我向一個馬來西亞的魚類專家問 "coral trout" 在其國家水域的出沒地點及背景資料,相信他不會明白我的問題。然而,假如我查詢 *Plectropomus leopardus* 的資料,他應能立即領會及提供答案。由此可見,任何物種的學名顯然對提高知識水平及促進國際科研合作有莫大幫助。

然而,對一本兒童圖書來說,生物的名字 (尤其是學名) 遠不及生物本身的資訊重要。我希望藉著這本書讓小朋友更加認識各種生物如何適應周圍的環境,及牠們在

the scientific names of any and all species are all too obviously essential for the advancement of knowledge and international scientific co-operation.

This is, however, a children's book and thus names, especially scientific ones, are not exactly of either use or importance. What is important are the creatures themselves because they impart not just them but how they are adapted to life on Earth and the roles they play in the wider ecology of life itself. And, thus, only a few words of information are provided about each of the species illustrated and these are mainly concerned with where it lives, its size and one interesting special feature about it.

Professor Brian Morton
2021

浩瀚的自然生態中扮演甚麼重要角色。因此，我在書中簡略介紹每一種動物的基本資料，包括牠們的居住地點、尺寸大小，及其他趣味小知識。

莫雅頓教授
2021

Scan the QR code to learn more about the sea creatures and animals in this book
掃描二維碼了解更多關於本書海洋及陸地動物的資訊

ARCTIC OCEAN
北極海

NORTH
AMERICA
北美

ATLANTIC OCEAN
大西洋

PACIFIC OCEAN
太平洋

SOUTH
AMERICA
南美洲

18

EUROPE
歐洲

ASIA
亞洲

AFRICA
非洲

INDIAN OCEAN
印度洋

AUSTRALIA
澳大利亞

Priacanthus tayenus

PURPLE-SPOTTED BIGEYE

The purple-spotted bigeye is a handsome, brilliant red fish, characterised by huge eyes, a tail with long filamentous tips and distinctive dark spots on the pelvic fins. It occurs throughout the Indo-West Pacific Oceans and is an important food fish.

長尾大眼鯛

長尾大眼鯛擁有迷人的紅色身體，明亮的雙目，尾巴兩端尖而長，腹鰭上有與別不同的黑點。牠遍布印度洋和西太平洋海域，屬於重要海產。

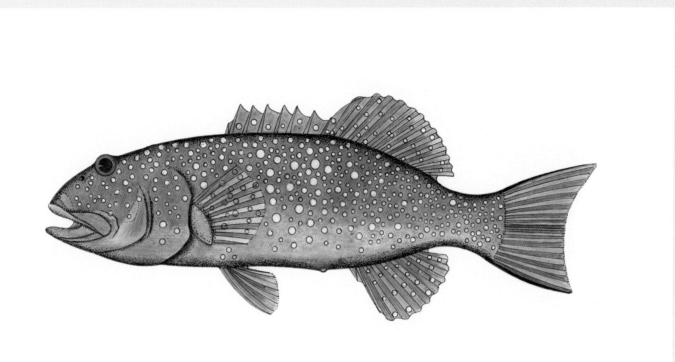

Plectropomus leopardus

CORAL TROUT

The coral trout is a spectacularly handsome grouper associated with coral reefs throughout the Indo-West Pacific, including Hong Kong. It grows to over one metre and lives for 20 years. Coral trout are ambush predators of smaller fish and crustaceans.

東星斑

東星斑是一種非常美麗的石斑,遍布印度洋和西太平洋海域的珊瑚礁,香港也可找到。牠身長可達一米多,可生存20年。東星斑是伏擊高手,喜歡吃較小的魚和甲殼類動物。

Psenopsis anomala

PACIFIC RUDDERFISH

The Pacific rudderfish (or melon seed or butterfish) lives in the northern part of the Western Pacific Ocean around Japan and Taiwan and within the East China Sea. It also occurs in the South China Sea close to Hong Kong.

瓜子鯧

瓜子鯧生活在西太平洋北部的日本，台灣和東海海域。牠亦會在靠近香港的南中國海聚居。

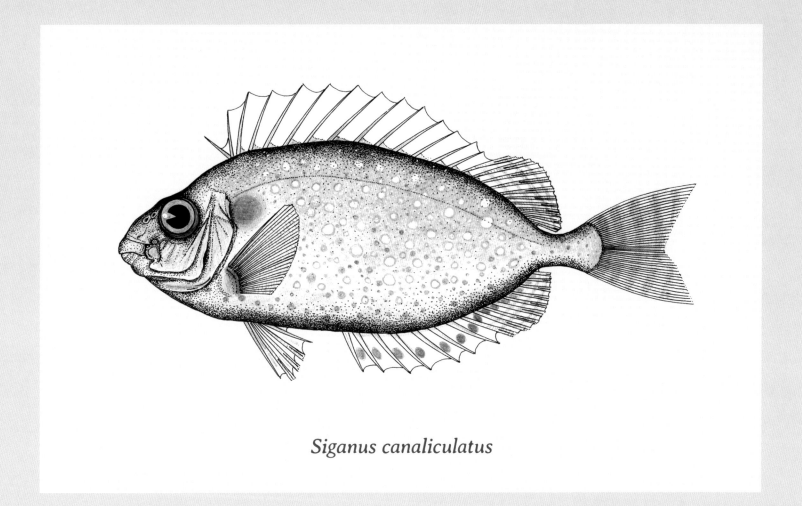

Siganus canaliculatus

WHITE-SPOTTED RABBITFISH

The white-spotted rabbitfish lives in river estuaries, lagoons and polluted harbours throughout the tropical Indo-West Pacific. It feeds on algae and reproduces virtually year-round, large females producing millions of eggs. Fin spines have venom glands that give a painful sting.

泥鯭

泥鯭生活在印度洋和西太平洋熱帶海域的河口、湖泊和受污染的港口。牠以藻類為食物,幾乎全年都在繁殖,較大的雌魚每次可排出幾百萬顆卵子。泥鯭的鰭棘有毒,被刺後會感痛楚。

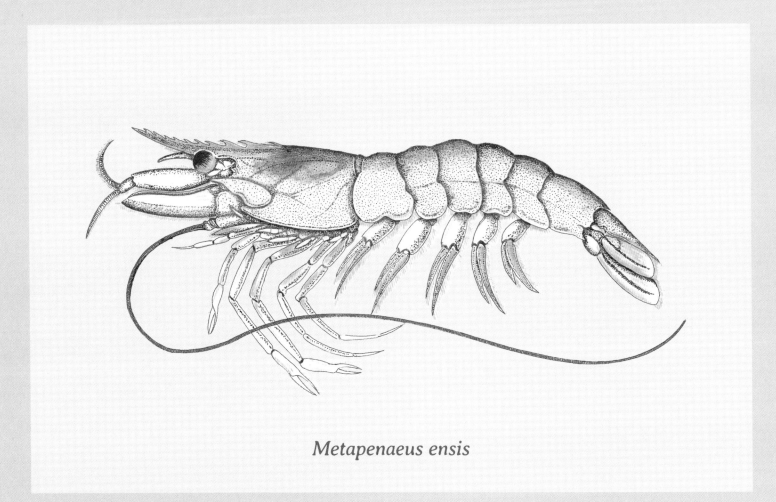

Metapenaeus ensis

GEI WAI SHRIMP

Gei wai shrimp is distributed throughout the Indo-West Pacific region. In the Asian element of its range it is cultured in coastal ponds. In Hong Kong this practice continues in the traditional gei wai's of the Mai Po Marshes Nature Reserve.

基圍蝦

基圍蝦分布於印度洋和西太平洋海域，亞洲則多於沿海池塘養殖。香港的米埔沼澤自然保護區至今保留這個養殖基圍蝦的傳統。

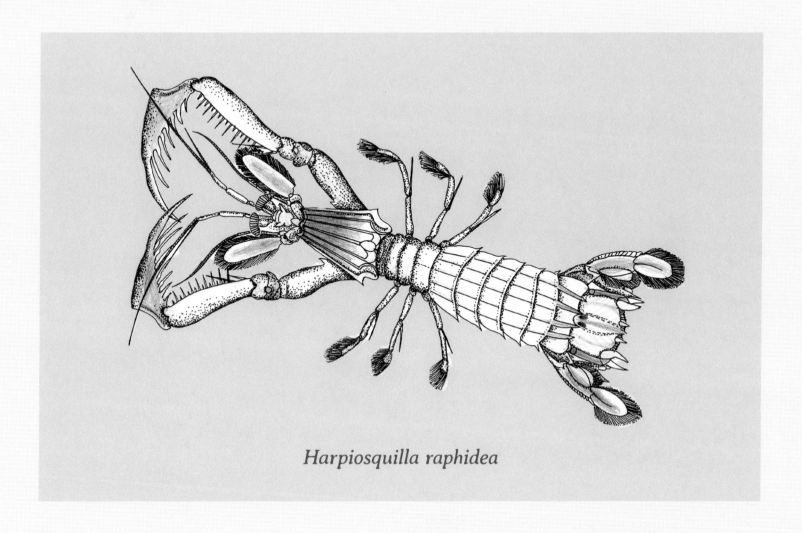

Harpiosquilla raphidea

MANTIS SHRIMP

Mantis shrimps are especially diverse in the tropical Indo-West Pacific including Hong Kong. One of the species, *Harpiosquilla raphidea,* can grow up to over 30 cm and has a pair of spiny chelae to stab and snag prey. It has large compound eyes — the most complex in the animal kingdom.

螳螂蝦

螳螂蝦的品種豐富，主要分布印度洋和西太平洋的熱帶海域，包括香港。其中一個品種棘突猛蝦蛄可生長至超過30厘米長，並擁有一對帶有長刺的螯足，有助捕捉獵物。牠有一對大複眼，在動物界稱得上最精密的眼睛。

Portunus pelagicus

BLUE SWIMMING CRAB

The blue swimming crab occurs throughout tropical Asia including Hong Kong, grows to a width of 20 cm and is an important fishery resource. Individual size and wild-caught captures have been declining of late so that this species is now widely cultured.

藍花蟹

藍花蟹遍布亞洲熱帶地區，香港也可找到。牠最闊可達20厘米，屬於重要海產。由於近年野外捕獲的藍花蟹體形越來越小，數目在下降，因此現在漁民已廣泛養殖這品種。

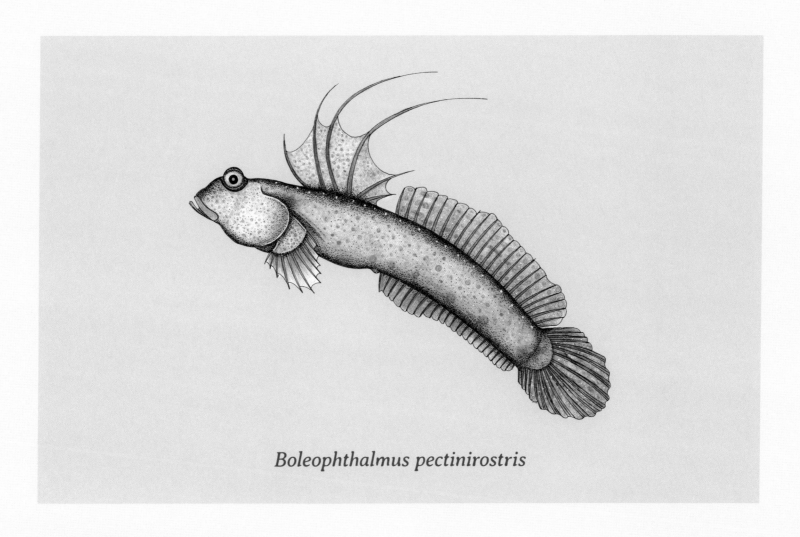

Boleophthalmus pectinirostris

GREAT BLUE SPOTTED MUDSKIPPER

The great blue spotted mudskipper is an air-breathing fish some 20 cm long living on seaward mudflats of mangrove stands in the Indo-West Pacific including Hong Kong. Mature males make upward leaps to demonstrate their ability to protect their territories and attract mates.

大彈塗魚

大彈塗魚是一種懂得呼吸空氣的魚，體長約20厘米，生活在印度洋和西太平洋海域包括香港沿岸紅樹林的泥灘。成年雄性經常飛彈而起，宣示領土，同時吸引異性。

Ailuropoda melanoleuca

GIANT PANDA

The giant panda lives in bamboo forests of the Tibetan Plateau. Uniquely, each hand has an extra opposing thumb. Giant pandas live for 20 years and females ovulate once each year giving birth to two cubs but only one survives.

大熊貓

大熊貓生活在青藏高原的竹林內。牠特別的地方，就是每隻手都有一隻呈相反方向的拇指。大熊貓壽命可達至20歲，雌性每年只會排卵一次，誕下兩隻小熊貓，但只有一隻會活下來。

Egretta Garzetta

LITTLE EGRET

The little egret is a small, 60 cm tall, member of the heron family Ardeidae. It lives throughout Eurasia in wetlands, marshes and along sheltered coastlines where they prowl shallow waters in search of prey food such as little fishes.

小白鷺

小白鷺體形細小,高約60厘米,屬於鷺科家族的成員。小白鷺遍布歐亞大陸的濕地、沼澤和擁有天然屏障的沿岸。牠們會在淺水地方尋找獵物,如水裡的小魚。

Lutra lutra

EURASIAN OTTER

The Eurasian otter is an aquatic, fresh water, mammal native to Eurasia. It has a long, slender, body and grows to one metre in length with the tail adding another third to that. Its diet mainly consists of fish.

歐亞水獺

歐亞水獺是歐亞大陸原生的淡水哺乳動物，身體細長，可達一米，尾巴有身體長度的三分之一。牠主要以魚類為食物。

Ailurus fulgens

RED PANDA

The red panda lives in southwestern China and is the size of a cat. It has a false thumb and feeds on bamboo like the giant panda. Despite their shared names, however, the red panda is unrelated to the giant panda.

喜馬拉雅小貓熊

喜馬拉雅小貓熊生活在中國西南部，體型與貓相似，並有一隻假拇指，跟大熊貓一樣喜歡吃竹葉。喜馬拉雅小貓熊與大熊貓的名字雖然很相似，但其實完全沒有關係。

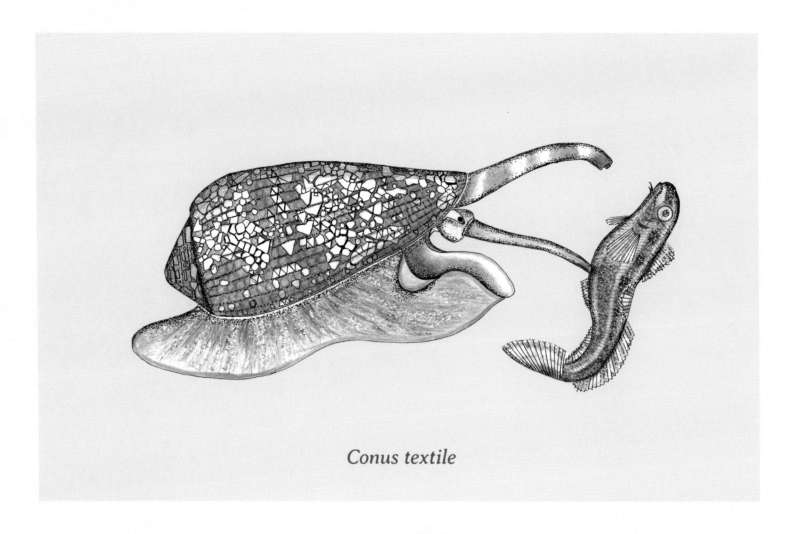

Conus textile

TEXTILE CONE

The textile cone occurs in the tropical Indo-Pacific region, including Hong Kong, living under large stones in shallow coastal waters. It attacks and feeds on other snails, worms and small fishes by harpooning and injecting them with a highly venomous toxin.

織錦芋螺

織錦芋螺生活在印度洋和太平洋的熱帶水域，香港也可找到，多棲息於沿海淺水的大石下面。牠以蝸牛、蠕蟲和小魚為食物，利用齒舌刺入獵物身體注入劇毒。

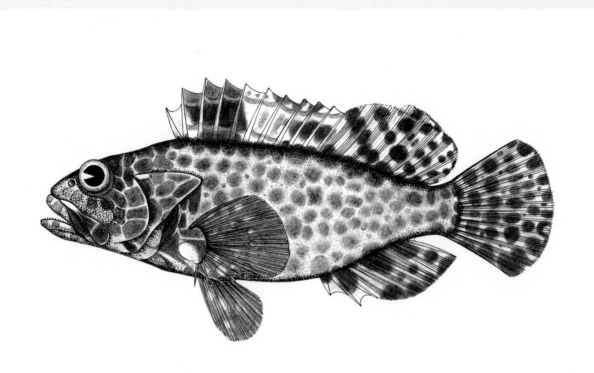

Epinephelus quoyanus

LONG-FINNED GROUPER

The long-finned grouper is a solitary, carnivorous, grouper living in the Western Pacific from Japan to Australia and throughout the waters of Southeast Asia, including Hong Kong. It attains a length of 40 cm and hunts its prey near the seabed.

花頭梅

花頭梅是一種獨居肉食性石斑,生活在日本至澳洲的西太平洋海域,遍布整個東南亞包括香港水域。牠的身長可達40厘米,喜歡在海底附近捕食獵物。

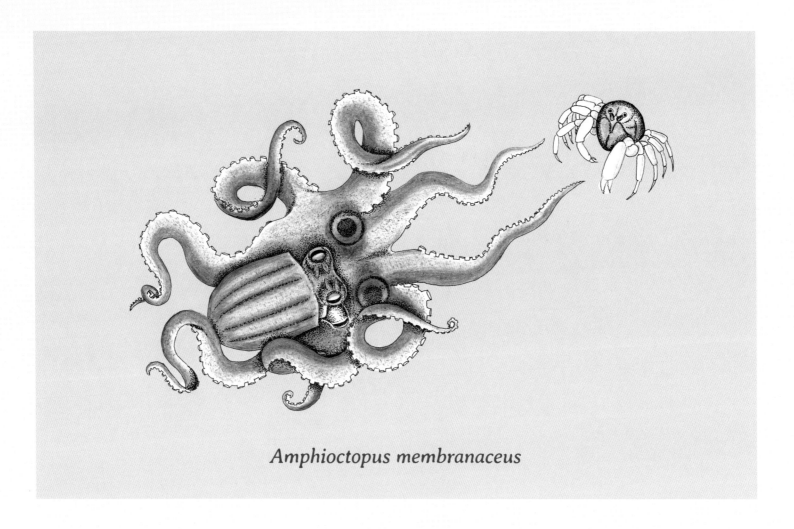

Amphioctopus membranaceus

WEBFOOT OCTOPUS

The webfoot octopus inhabits the tropical waters of the Indian and Pacific Oceans, including Hong Kong and lives in empty clam shells on shallow beaches. It creates a pair of huge fake eyes when hunting food, especially crabs, and when frightened.

短爪章魚

短爪章魚生活在印度洋和太平洋包括香港的熱帶水域,多棲息於淺灘的空蛤殼內。牠特別喜歡吃螃蟹,捕食或遇險時會假裝有一對大眼。

Sepioloidea lineolata

STRIPED DUMPLING SQUID

The striped dumpling squid lives in shallow, sandy, sea grass beds off southern Australia. It is only 6 cm long. It hides within the sand and waits for unwary shrimps, crabs and fishes to wander by, which it pounces on and devours.

條紋睡衣魷魚

條紋睡衣魷魚生活在澳洲南部,棲息於沙質和長有海草的淺水海床。牠只有六厘米長,喜歡躲藏在沙裡,等待毫無防避的海蝦、螃蟹和小魚游過,然後撲出來獵吃。

GHOST JELLYFISH

Every summer as South China Sea waters invade Hong Kong, the ghost jellyfish arrives with them, in large numbers. Its bell can grow to 50 cm in diameter and its tentacles are 10 metres long. Barnacles and fishes live with it.

幽靈水母

幽靈水母在每年夏天都會隨著南中國海的水流進入香港，數量驚人。牠的身體直徑可達50厘米長，觸鬚更可長達十米，會與甲殼動物和魚類一同生活。

Cyanea nozakii

ATLANTIC PUFFIN

The Atlantic puffin lives in the northern Atlantic Ocean. The colourful bill, waddling gait and behaviour give it nicknames — "clown of the sea" and "sea parrot". Puffins nest in clifftop colonies, digging a burrow in which one chick is reared.

This species has been rated in 2015 as vulnerable by the International Union for Conservation of Nature (IUCN).

北極海鸚

北極海鸚生活在北大西洋,鳥嘴色彩豐富,走路時搖搖擺擺,因此贏得「海上小丑」和「海鸚鵡」的外號。牠喜歡在懸崖頂尋找洞穴築巢,在巢內撫養一隻小寶寶成長。

該物種在2015年被國際自然保護聯盟評為易危物種。

Fratercula arctica

SCALLOPED SPINY LOBSTER

The scalloped spiny lobster lives along the coasts of the Indian and Pacific Oceans. It typically grows to a length of 30 cm and is abundant on coral and fringing rocky reefs. It is highly social, preferring to congregate in groups.

波紋龍蝦

波紋龍蝦生活在印度洋和太平洋沿岸，通常長達30厘米，大多生活在珊瑚和外圍的礁石上。牠很喜歡群居，經常成群結隊。

Panulirus homarus

TOMPOT BLENNY

The tompot blenny occurs in the waters of Western Europe from southern England to Morocco. It also occurs throughout the Mediterranean. At 30 cm in length, it is a large blenny that lives in rock cavities. Males defend their territories aggressively.

淺紅副䲁

淺紅副䲁生活在西歐一帶海域，從英格蘭南部至摩洛哥都可找到，還遍布整個地中海。淺紅副䲁身長30厘米，屬於體型較大的副䲁品種，棲息於岩石洞內，雄性會勇猛捍衛領土。

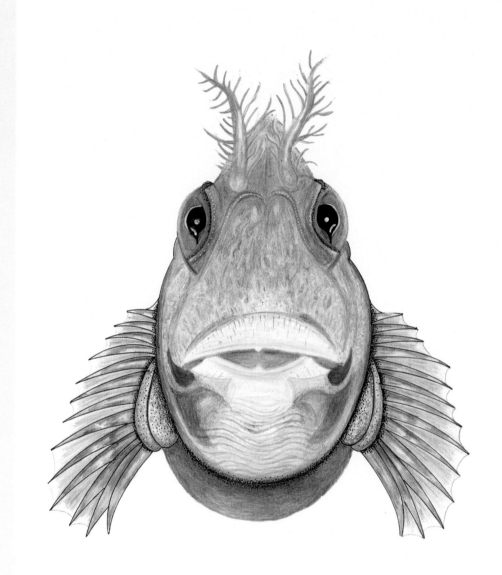

Parablennius gattorugine

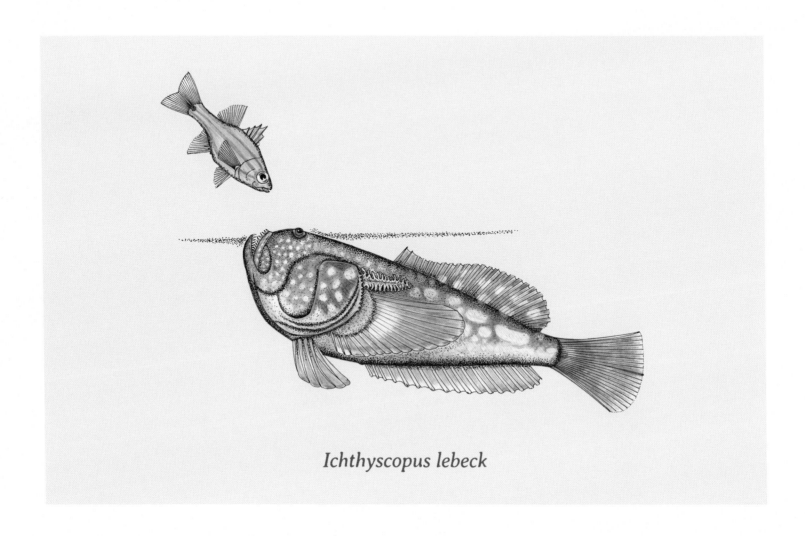

Ichthyscopus lebeck

STARGAZER

The stargazer is known from only a few locations, including Hong Kong. This bizarre fish lies buried in shallow sands with only tiny eyes sticking out, staring skyward. Any inquisitive fish attracted to it is sucked into its enormous mouth.

瞻星魚

香港是瞻星魚少有的棲息地之一。這種外形特別的魚喜歡埋在淺沙中，只露出一對向上望的小眼睛。任何魚只要好奇游過去，就會被牠張開大口吃掉。

Sparisoma cretense

MEDITERRANEAN PARROTFISH

The Mediterranean parrotfish occurs along rocky coastlines in the Mediterranean and the eastern Atlantic. It nibbles algae off the rocks. Females are red and yellow while males are greyish overall. Individuals are firstly female and then change sex to become male.

異齒鸚鯛

異齒鸚鯛生活在地中海和東大西洋沿海的岩石，以岩石上的藻類為食物。雌性身體又紅又黃，雄性則全身灰色。最特別的地方，就是異齒鸚鯛初時都是雌性，然後變為雄性。

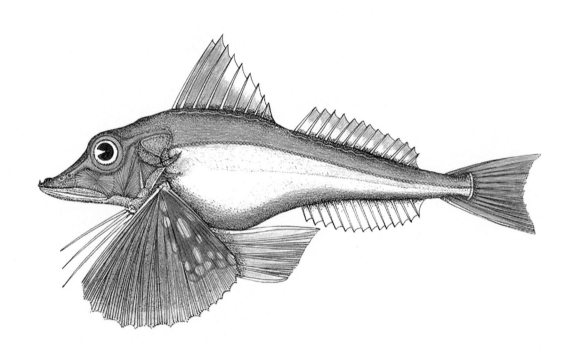

Chelidonichthys lucerna

TUB GURNARD

The tub gurnard has a large armoured head and a long, tapering, body. The pectoral fins have three finger-like rays used for "walking" along the seabed. The remainder of the pectoral fins are enlarged and used to "fly" away from danger.

細鱗綠鰭魚

細鱗綠鰭魚的頭部很大,像一個裝甲,長長的身體向尾部收窄。胸鰭有三根形似手指細爪,可以沿著海床「行走」。牠遇到危險時,胸鰭會張大,然後像「飛」一樣逃走。

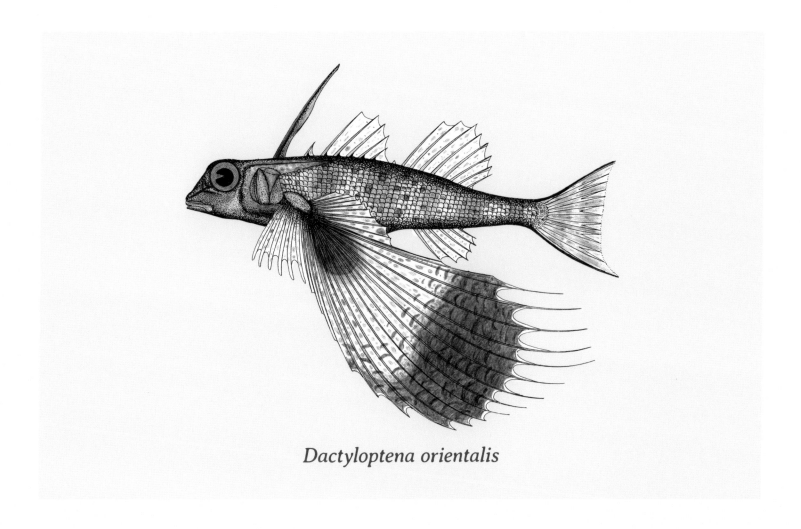

Dactyloptena orientalis

ORIENTAL FLYING GURNARD

Oriental flying gurnard grows to 40 cm in length and lives throughout the Indo-West Pacific region, including Hong Kong. The wing-like pectoral fins are brightly coloured and their first six rays are modified into feelers for "walking" over the seabed.

東方豹魴鮄

東方豹魴鮄身長可達40厘米，生活在印度洋和西太平洋海域，在香港也可找到。牠的胸鰭貌似翅膀，色彩鮮豔，最前的六根鰭條媲美觸鬚，可在海床上「行走」。

Phoca vitulina

HARBOUR SEAL

The harbour seal lives along the northern coasts of the Atlantic and Pacific Oceans. Adult males can be two metres long and weigh 200 kilogrammes. Females give birth once each year to a single pup, which they care for alone.

港海豹

港海豹生活在大西洋和太平洋的北部海岸。成年雄性可長約兩米，重200公斤。雌性港海豹每年誕下一個小寶寶，然後獨力照顧。

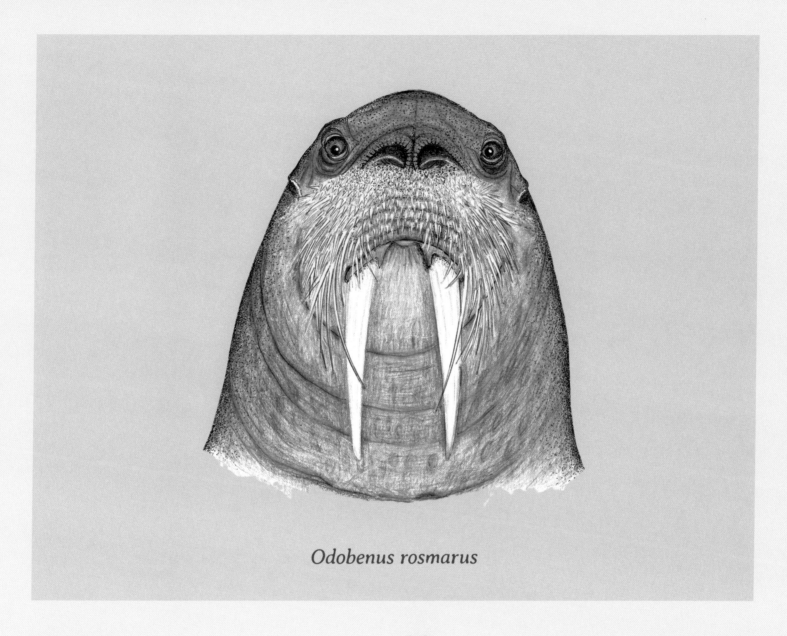

Odobenus rosmarus

WALRUS

Two sub-species of walrus occur in the Atlantic and Pacific. Adults weigh one tonne and have huge canine tusks. Walruses herd peacefully except when males fight for females. Females give birth distant from the crowd to stop calves being crushed.

海象

海象有兩個亞品種，生活在大西洋和太平洋。成年海象重達一噸，有一對很大的象牙。海象彼此和平相處，只有雄性在求偶時才會大打出手。雌性海象生產時會到僻靜的地方，避免小寶寶受到傷害。

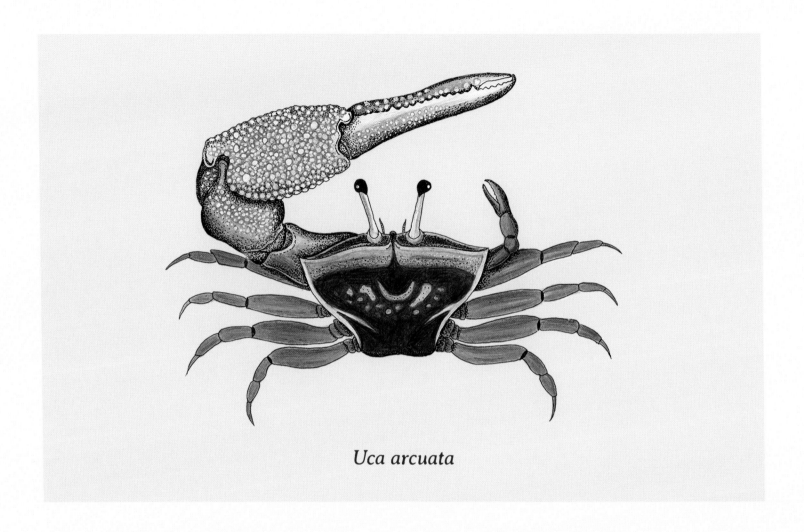

Uca arcuata

BOWED FIDDLER CRAB

The bowed fiddler crab lives throughout the tropics, including Hong Kong, and inhabits inter-tidal muds seaward of mangrove stands. Males (with a carapace width of 25 mm) have a large chela for display and fighting and a smaller feeding one.

弧邊招潮蟹

弧邊招潮蟹遍布整個熱帶地區，香港也可找到，生活在沿海的紅樹林泥灘中。雄性的甲殼闊度約25毫米，有一個大蟹鉗，可用來宣示實力和打鬥，而較小的蟹鉗用來進食。

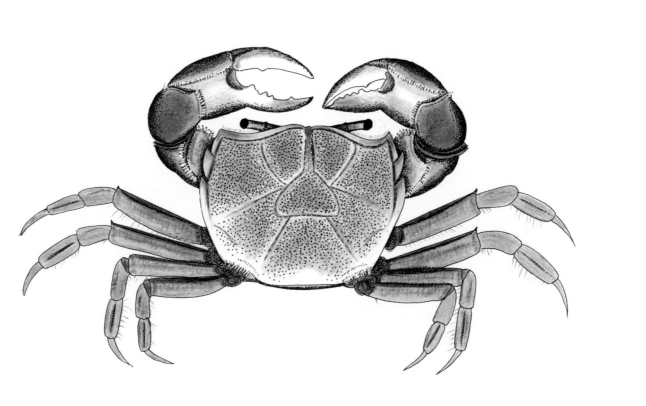

Chasmagnathus convexus

MANGROVE CRAB

Mangrove crab is endemic to East Asia and inhabits deep burrows within mangrove stands. The carapace is 5 cm across. It is common in Hong Kong's mangroves and emerges at night to feed on sediments, macroalgae and mangrove leaf litter.

隆背張口蟹

隆背張口蟹是東亞特有品種，生活在紅樹林的泥沼內，甲殼闊五厘米。牠在香港的紅樹林也很常見，晚上會出來活動，喜歡吃沉積物、大型藻類和紅樹的落葉。

Sebastiscus marmoratus

SCORPIONFISH

The scorpionfish occurs in the Western Pacific from Japan south to the Philippines and including Hong Kong. It is colour camouflaged to resemble its chosen habitat. The fins are large with the dorsal one, in particular, possessing venomous spines.

石狗公

石狗公生活在日本南部至菲律賓的西太平洋海域，香港也可發現。石狗公身體有保護色，顏色與生活環境相近。牠的鰭很大，尤其是背鰭更有毒刺。

Histrio histrio

SARGASSUM FISH

The sargassum fish is an ambush predator of small creatures that inhabit the fronds of tropical sargassum seaweeds. It is covered with numerous feathery flaps of skin and undergoes rapid colour changes so that it is camouflaged within the algal mats.

裸躄魚

裸躄魚是埋伏高手，專門捕捉生活在熱帶馬尾藻葉中的小生物。牠身上長有羽毛狀的皮，可以迅速改變顏色，隱藏在海藻羣內。

Takifugu alboplumbeus

PUFFER FISH

Takifugu alboplumbeus is a species of puffer fish which occurs in Japan and the coastline of China, including Hong Kong. It attains a length of 23 cm and can inflate itself by filling its stomach with water. Each fish contains lethal amounts of a neurotoxin called tetrodotoxin.

河豚

鉛點東方魨是生活在日本和中國沿海,包括香港的河豚。牠體長可達23厘米,可將大量海水灌入胃內使身體膨脹。河豚含有河豚毒素,屬於一種神經劇毒,足以致命。

Pygocentrus nattereri

RED-BELLIED PIRANHA

The red-bellied piranha, half a metre in length, is distributed throughout South America living in rivers such as the Amazon. It occurs in shoals and has a ferocious reputation, even attacking humans. They are, however, mainly scavengers, fulfilling a clean-up role.

紅腹食人魚

紅腹食人魚身長半米,遍布南美洲,例如亞馬遜河流域。紅腹食人魚會在淺灘出沒,生性兇猛,甚至會襲擊人類。不過牠主要的食物是其他生物的屍體,在水裡扮演清道夫的角色。

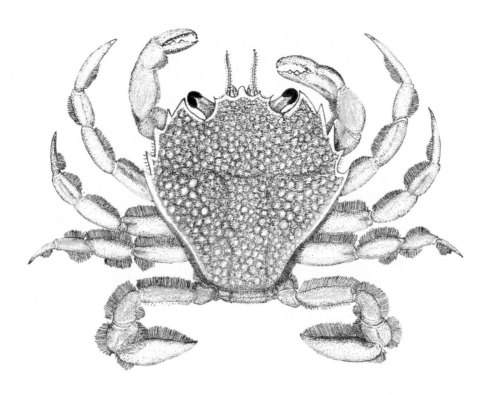

Portumnus latipes

PENNANT'S SWIMMING CRAB

Pennant's swimming crab occupies the sub-tropical shorelines of the Mediterranean and, increasingly, the North Sea perhaps due to global warming. Its rear limbs are paddle-like for swimming and burrowing. It is omnivorous but favours surf clams, which share its habitat.

細長梭子蟹

細長梭子蟹遍布地中海的亞熱帶海岸，可能由於全球暖化，生活在北海的數目逐漸增多。牠的後肢形似船槳，適合游泳和挖洞。細長梭子蟹雖然是雜食，但最愛吃與其生活在同一環境的北寄貝。

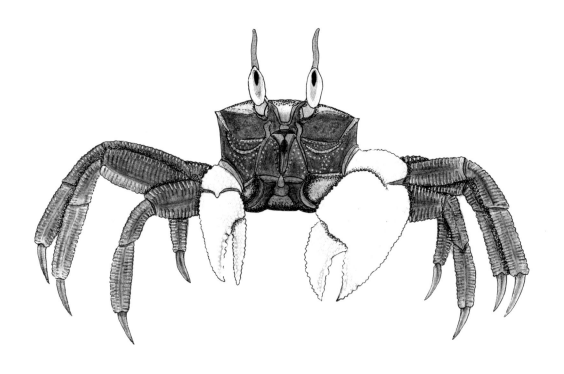

Ocypode ceratophthalmus

LONG-HORNED GHOST CRAB

The long-horned ghost crab inhabits the Indo-West Pacific region, including Hong Kong. Up to 8 cm across, it builds burrows towards the rear of sandy beaches and feeds on other creatures including turtle hatchlings. It runs extremely fast to catch its prey.

角眼沙蟹

角眼沙蟹生活在印度洋和西太平洋海域，在香港也能找到。牠的闊度可達八厘米，喜歡在沙灘近陸地方向挖洞穴。角眼沙蟹能極速飛跑，捕捉獵物，包括剛孵化出來的小龜和其他生物。

SEA MOTH

Pegasus is named after the winged horse of Greek mythology. The four species of sea moth are all characterised by wing-like pectoral fins, a long, spiny nose and large eyes. They occur in deeper waters throughout the Indo-West Pacific Ocean.

海蛾魚

海蛾魚的拉丁學名 Pegasus 取自希臘神話中的有翼飛馬。海蛾魚的四個品種均有形似翅膀的胸鰭，鼻子長而有刺，眼睛很大。牠們生活在印度洋和西太平洋的深海。

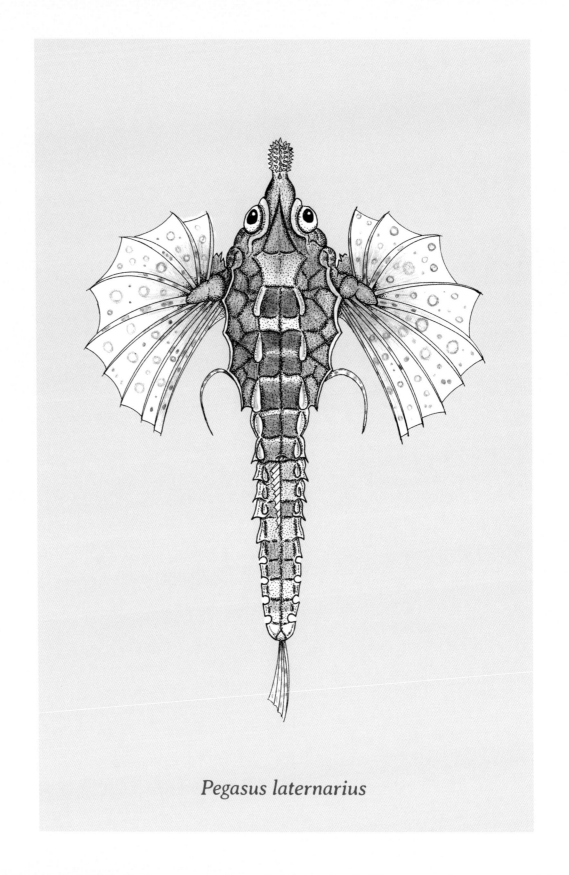

Pegasus laternarius

YELLOW SEAHORSE

The yellow seahorse lives in the tropical waters of the Indo-West Pacific. After courtship, males and females intertwine their tails. Belly to belly, the female inserts her eggs into the male's pouch where they are fertilized and nourished until birth.

This species is listed in the Convention on International Trade in Endangered Species of Wild Fauna and Flora (CITES).

管海馬

管海馬生活在印度洋和西太平洋的熱帶水域。交配時，雄性和雌性管海馬尾巴相交，腹部對著腹部，雌性將卵子注入雄性的孵育袋內，受精卵會在裡面成長，直至小海馬出生。

該物種已被列入瀕危野生動植物種國際貿易公約。

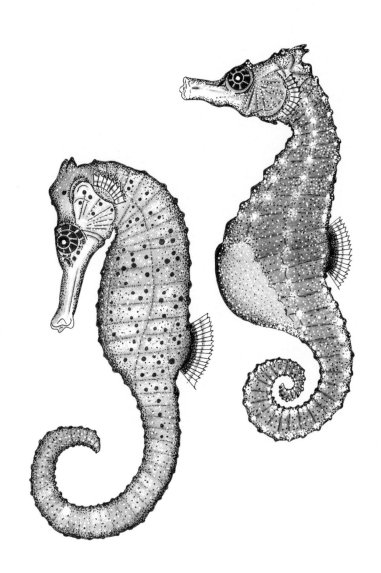

Hippocampus kuda

Luidia maculata

EIGHT-ARMED SAND STAR

The eight-armed sand star is a tropical carnivorous sea star, which occurs throughout the Indo-West Pacific. Unlike most sand stars, however, with five arms, it has eight. Species of Luidia are also different, moreover, because their larvae can asexually divide by cloning.

斑砂海星

斑砂海星是熱帶肉食性海星,遍布印度洋和西太平洋海域。大多數海星只有五臂,但斑砂海星卻有八臂。砂海星屬的品種另外一個獨特之處,是幼蟲可進行無性繁殖,自行複製出更多幼蟲。

Archaster typicus

SAND STAR

The sand star grows to 15 cm in diameter and inhabits sheltered sandy beaches throughout the Indo-West Pacific, including Hong Kong. Unusually, the sand star engages in pseudocopulation whereby males climb atop females and they release gametes or external fertilisation simultaneously.

飛白楓海星

飛白楓海星直徑可長至15厘米，生活在印度洋和西太平洋海域風浪較少的沙灘，香港也可發現。飛白楓海星最特別的地方，就是會進行假交配，雄性爬到雌性身上，然後兩者同時釋放精子和卵子進行體外受精。

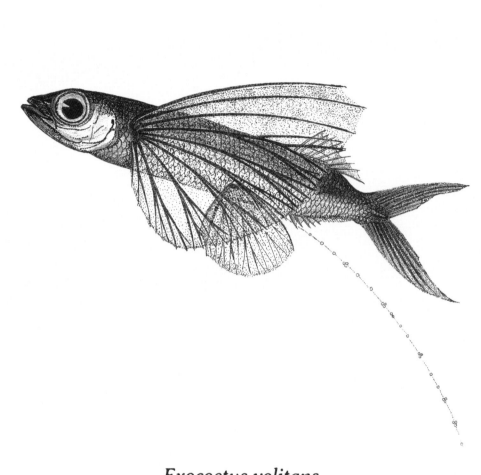

Exocoetus volitans

TROPICAL TWO-WINGED FLYING FISH

Flying fishes occupy tropical, oceanic, waters worldwide. One of the species, *Exocoetus volitans*, has a maximum length of 30 cm and a streamlined body iridescent blue above and silvery white below. It can glide above the sea to escape predators for 100 metres.

大頭飛魚

飛魚遍布全球的熱帶海域，其中一個品種是大頭飛魚，最長可達30厘米，擁有流線形的身體，魚背呈閃耀藍色，魚腹呈銀白色。牠可以在海面上滑行100米逃避敵人。

Ursus maritimus

ICE BEAR

The ice bear is an Arctic apex predator eating seals and the carcasses of dead mammals. The largest male boar ever recorded weighed over one tonne and stood to a height of 3.5 metres. A female sow weighs half that.

北極熊

北極熊是北極最兇猛的食肉動物,喜歡吃海豹和哺乳類動物的屍體。有記錄以來最大一頭雄性北極熊重達一噸,身高3.5米。雌性體重則是雄性的一半。

Ocean Park Conservation Foundation, Hong Kong ("OPCFHK") is committed to working together with people from all walks of life to promote ecological conservation across Asia through collaborative fundraising, scientific research, capacity building, and education. With a particular focus on marine conservation, OPCFHK continuously strives to tackle the threat and impact of habitat destruction and fragmentation, pollution, overfishing, climate change, and the illegal wildlife trade. OPCFHK has allocated funds totalling more than HK$98.62 million to over 500 research studies in pursuit of these goals. For example, it has established a dedicated Cetacean Stranding Response Team and extended its scope of work to include endangered sharks and other finfish. OPCFHK remains as committed as ever to inspiring students of all ages about the importance of conservation and research through its numerous programmes, initiatives, and public campaigns. The recent No Straw Campaign aimed at encouraging the people of Hong Kong to stop using plastic drinking straws and other disposable plastic products — a key part of marine conservation.

Indeed, no one has worked as tirelessly or been as dedicated to the protection of Hong Kong's marine ecology as the former Chairman of OPCFHK, Professor Brian Morton.

香港海洋公園保育基金(OPCFHK)致力與各界人士攜手合作,通過協作籌款、科學研究、能力建設和教育,推動亞洲地區的生態保育。保育基金特別關注海洋保育,不斷努力應對生態環境破壞和破碎化、污染、過度捕撈、氣候變化及非法野生動物貿易的威脅和影響。為實現這些目標,保育基金已撥款展開超過500多個研究項目,總額超過9,862萬港元。例如,保育基金成立了一個專門的鯨豚擱淺行動小組,並將其工作範圍擴大到包括瀕危的鯊魚和其他有鰭魚類。保育基金一如既往地致力於通過不同的項目和計劃、倡議及公眾活動,鼓勵不同年齡層的學生瞭解保育和研究的重要性。最近的「無飲管運動」旨在鼓勵香港人停止使用塑膠飲管和其他一次性塑膠製品 — 這是海洋保育的重要一環。

事實上,保育基金前主席莫雅頓教授在保護香港海洋生態方面的工作貢獻良多,他那孜孜不倦和盡心盡力的堅持可謂沒人能及。莫雅頓教授於1969年出任香港大學動物學講師,畢生致力於海洋生物研究。在香港工作的34年裏,他為保護這個美麗群島

Having taken up the position of Lecturer of Zoology at The University of Hong Kong in 1969, Professor Morton dedicated his life to marine biological research. During his 34 years in Hong Kong, he arguably did more to preserve the unique biodiversity of this beautiful archipelago than anyone. Among his countless achievements, Professor Morton's work led to the passing of The Marine Parks Ordinance Act of 1995, and he was also the founding director and visionary creator of The Swire Institute of Marine Science. A multiple award-winning, internationally renowned scientist, Professor Morton leaves an incredible legacy not only in terms of his work but also in his mentoring of successive generations of ecologists and conservationists. Even in retirement, he was as active as ever, providing forty of his drawings to The Fullerton Ocean Park Hotel Hong Kong for creating this marine conservation book, with all proceeds from selling of the book going to OPCFHK. A truly great scientist who inspired everyone he met on the importance of marine conservation, Professor Morton has now passed the torch that he ignited so many years ago to the next generation of ecologists and conservationists. OPCFHK would like to take this opportunity to express its sincere gratitude for his decades of dedicated service.

獨特的生物多樣性所付出的努力可以說比任何人多。莫雅頓教授成就無數，他促成了在1995年通過的《海岸公園條例》，亦是太古海洋科學研究所的創始所長。作為一名屢獲殊榮的國際知名科學家，莫雅頓教授不僅在學術研究上影響深遠，亦培養了幾代生態學家和自然保育專家，啓發無數的學生。即使在退休後，他仍一如既往地積極工作，為新落成的香港富麗敦海洋公園酒店提供了40幅自己的畫作，製作成此海洋保育書籍，而所有收益將撥捐香港海洋公園保育基金。莫雅頓教授是一位真正偉大的科學家，他激勵著他遇到的每一個人，讓他們認識到海洋保育的重要性，現在他已經把多年前點燃的火炬傳給了下一代的生態學家和生態環境保育者。保育基金謹藉此機會對他數十年的熱誠服務表達由衷的感謝。

 You can contribute to conserving Hong Kong's marine environment by donating to OPCFHK.
www.opcf.org.hk

 您亦可以通過向香港海洋公園保育基金捐款，為保護香港的海洋環境做出貢獻。
www.opcf.org.hk

Professor Brian Morton
莫雅頓教授

(1942 - 2021)

1940s

1960s

Professor Brian Morton was a world-renowned marine ecologist. Born in August 1942 during World War II in an Anderson air raid shelter. He obtained his degree from King's College, the University of London, and later taught Ecology and Biodiversity at the University of Hong Kong from the early 1970s until his retirement in 2003.

A pioneer in Hong Kong's marine conservation efforts, Professor Morton was a founding member of the World Wide Fund for Nature Hong Kong (WWFHK), and also contributed to the formation of the Mai Po Nature Reserve, Swire Institute of Marine Science (SWIMS), Hoi Ha Wan Marine Reserve, as well as the Marine Biological Association of Hong Kong. He was also Chairman of the Board of Trustees of

莫雅頓教授是國際知名的海洋生態學家，於1942年8月在第二次世界大戰期間出生於一個防空洞。他持有倫敦大學倫敦國王學院所頒發的學位，自1970年代起於香港大學生態學及生物多樣性學系任教，至2003年榮休。

作為引領保育香港海洋環境的先導，莫雅頓教授是世界自然基金會香港分會創會會員，多年來促成米埔自然保護區、太古海洋科學研究所、海下灣海洋生物中心、以及成立香港海洋生物學協會等。他亦曾任2001年至2003年海洋公園保育基金的董事會主席。

The Ocean Park Conservation Foundation (OCPF) between 2001 and 2003.

Professor Morton was a prolific writer having published over 25 books and 600 scientific papers. He was invested as Knight (Ridder) in the Order of the Golden Ark, The Netherlands, and the Order of the British Empire, U.K and received the Duke of Edinburgh WWF Gold Medal for his work in marine conservation.

Upon his retirement, he returned to his childhood home in Littlehampton on the south coast of England, where his love of marine ecology started and there, he continued his research, book writing and drawing.

莫雅頓教授出版的書籍超過25本，科學論文和報告逾600篇。他曾獲頒荷蘭金色方舟騎士勳章、英國的OBE勳銜及愛丁堡公爵環境保育勳章等，表揚他對海洋生態保育的貢獻。

莫教授於退休後回到位於英國南海岸，始於對海洋生態熱愛的童年故居Littlehampton，繼續他的研究工作、寫作和繪畫。

DID YOU KNOW?
你知道嗎?

_____

DID YOU KNOW?

With 84 hard coral species, Hong Kong has a higher coral diversity than the Caribbean Sea.

DID YOU KNOW?

Hong Kong has 6 marine parks and one marine reserve.

你知道嗎?

香港有84種石珊瑚,珊瑚的多樣性比加勒比海還要多。

你知道嗎?

香港有六個海岸公園及一個海岸保護區。

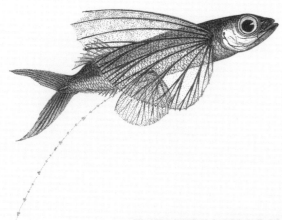

DID YOU KNOW?

Sha Chau and Lung Kwu Chau Marine Park is the largest marine park in Hong Kong. It occupies 1200 hectares of open water, and the size is about 60 times the size of Victoria Park.

DID YOU KNOW?

A total of 997 marine fish species have been recorded in Hong Kong. This is about 30% of the total fish found in the South China Sea.

你知道嗎?

沙洲及龍鼓洲海岸公園是香港最大的海岸公園。它的海域面積達1200公頃，大約是維多利亞公園的60倍面積。

你知道嗎?

香港錄得的海水魚達997種，約佔南中國海海魚種類的三成。

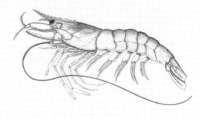

ACKNOWLEDGEMENTS
鳴謝

———————

As the first green-certified waterfront resort in Hong Kong, The Fullerton Ocean Park Hotel Hong Kong is committed to marine education and conservation in our city. Accredited with WELL Precertification under the WELL Building Standard™ v2 and aiming to achieve Final Gold rating under BEAM Plus New Buildings V1.2, we strive to engage in sustainable practices with our commitment to sourcing 100-percent sustainable seafood.

It has been our great honour to have collaborated with "Hong Kong's father of marine conservation", the late Professor Brian Morton, on this meaningful bilingual children's guide. Professor Morton's beautiful hand-drawn illustrations bring to life forty fascinating sea creatures and animals from Hong Kong and around the world.

Professor Morton's departure in March 2021 marks a great loss to the world of marine biology. We take this opportunity to express our sincere gratitude to the beloved

作爲香港首間獲得綠色認證的臨海渡假村，香港富麗敦海洋公園酒店一直致力在香港推廣海洋教育和保育。我們榮獲 International WELL Building Institute™ (IWBI™) 頒發《WELL 建築標準™》v2 預認證，並以取得綠建環評新建建築 1.2 版金級認證為目標，積極參與可持續實踐，履行採購 100% 可持續海鮮的承諾。

我們十分榮幸能與「香港海洋生態保育之父」—已故莫雅頓教授合作，製作這本別具意義的雙語兒童圖書。莫雅頓教授親手繪畫了 40 種來自香港及世界各地的海洋生物與動物，無不精緻美麗、栩栩如生。

莫雅頓教授於 2021 年 3 月與世長辭，乃海洋生物學領域的重大損失。我們在此向這位敬愛的海洋生態學家致以誠摯的謝意，感謝他啟發了我們對海洋環境的關愛、尊重和保護。

marine ecologist for inspiring love, respect and protection for our marine environment.

We are also very grateful to Professor Morton's daughter, Bryony, for completing this book with us after the passing of her father. It is our hope that the publication of this book will help to further Professor Morton's legacy and celebrate his inspirational work.

Very special thanks to Dr Leung Siu-fai and Mr Lau Ming-wai for their enthusiastic support towards this project and for their contributions to the Foreword and Message in the book.

To Ocean Park Conservation Foundation, Hong Kong (OPCFHK), who will receive the book proceeds, thank you for your wonderful work and warm support for this initiative.

The Fullerton Ocean Park Hotel Hong Kong

我們非常感謝莫雅頓教授的女兒，Bryony在她父親離世後與我們一起完成了這本書。我們希望這本書的出版能傳承莫雅頓教授的非凡成就，並紀念他鼓舞人心的工作成果。

我們亦十分感激梁肇輝博士和劉鳴煒先生對這個項目的熱情支持，以及為這本書分別撰寫了前言和香港海洋公園主席的話。

本書的所有收益，將全數撥捐予香港海洋公園保育基金。我們衷心感謝保育基金的出色工作和對這個項目的熱心支持。

香港富麗敦海洋公園酒店